Complex Contours, Conformal Constructions

A detailed guide to conformal mappings, boundary value problems, and applications in physics and engineering

Forrest Herrera

Table of Contents

1 Introduction to Conformal Mappings

Conformal mappings are a powerful tool in complex analysis, offering elegant solutions to problems that would otherwise be intractable. Their defining characteristic – the preservation of angles – makes them particularly useful in modeling physical phenomena governed by Laplace's equation, such as fluid flow and electrostatics. This preservation of angles, a local property, ensures that the structure of curves intersecting at a point remains unchanged after the transformation, providing a geometric intuition often lacking in more abstract mathematical methods. Understanding this fundamental property is crucial for appreciating the broader applications of conformal mappings.

The foundation for studying conformal mappings rests firmly on the understanding of complex numbers and functions. Complex numbers, with their real and imaginary parts, form the domain on which these mappings operate. Functions of a complex variable, capable of mapping one complex plane to another, provide the mechanism for the transformations. We will explore these functions, their derivatives, and their behavior in the complex plane, establishing a crucial link between the analytic properties of these functions and the geometric properties of the resulting mappings. A thorough grasp of these concepts is fundamental to navigating the more advanced applications later in this book.

Now let's move to the core concept: conformal transformations themselves. A conformal mapping is a function that preserves angles locally. Imagine drawing two curves that intersect at a point in one plane. A conformal map transforms these curves into two new curves in a second plane; the angle between the original curves at their point of intersection will be identical to the angle between the transformed curves at their corresponding intersection point. This property is what makes conformal mappings so useful, allowing us to simplify complicated geometries into more manageable ones.

While the angle-preserving property is local, its global consequences are far-reaching. For example, in fluid dynamics, the streamlines of a fluid flow, which intersect orthogonally to the equipotential lines, maintain their orthogonality under conformal mapping. This allows us to transform complicated flow patterns in irregular geometries into simpler, more readily analyzable ones in the transformed plane. This simplicity simplifies calculations significantly and allows us to find solutions for challenging problems in a methodical way.

One particularly useful tool in visualizing and understanding conformal mappings is the use of color plots. By assigning colors to different values of a complex function, we can visualize how the mapping transforms regions in the complex plane, and how these transformations impact the geometry of the curves being mapped. These color maps provide a valuable aid both for conceptual understanding and for identifying potential issues in computations. This method is not only instructive but is also widely used in computational analysis and simulation.

The next chapter delves into specific elementary conformal maps–M\"obius transformations, exponential and logarithmic maps, and power and root functions. These form the building blocks for constructing more complex mappings. Mastery of these elementary transformations is essential for any deeper work with conformal mappings. They provide a starting point for the exploration of advanced topics and the development of a more comprehensive skill set.

The utility of conformal mapping extends far beyond theoretical considerations. Their application in solving practical problems in diverse fields, such as engineering and physics, underscores their significance. The following chapters will explore these applications in detail. Through carefully chosen examples, we will demonstrate the power and elegance of the method and further solidify understanding of the underlying principles.

Understanding conformal mappings unlocks a profound ability to simplify and solve problems previously considered too complex. The elegance of the approach combines geometric intuition with the power of complex analysis, leading to efficient solutions of many types of engineering problems. Therefore, mastering these methods will prove invaluable to anyone seeking to solve mathematical and physical problems related to fields and flows. By mastering conformal mappings, we gain access to a sophisticated and effective arsenal for tackling a wide array of complex tasks and challenges. The remaining chapters of this book will equip you with the skills to apply these tools effectively in your own work.

1.1 Complex Numbers & Functions

Complex numbers, represented as $z = x + iy$, where 'i' is the imaginary unit ($i^2 = -1$), form the foundation of our exploration. Understanding their graphical representation in the complex plane, with x as the real axis and y as the imaginary axis, is crucial. Operations like addition, subtraction, multiplication, and division are straightforward extensions of real number arithmetic, but with the added nuance of imaginary components. The magnitude (or modulus) of a complex number, denoted by $|z|$, represents its distance from the origin, while the argument (or phase) represents its angle with respect to the positive real axis. These concepts underpin many of the transformations we will explore later. Mastering complex arithmetic is the first step toward understanding the rich world of conformal mapping.

Functions of a complex variable, denoted as $w = f(z)$, map points in the complex z-plane to points in the complex w-plane. These functions aren't merely transformations of coordinates; they possess unique properties tied to their analyticity, a property indicating their differentiability within a region. An analytic function possesses a derivative at every point within a region, ensuring its smooth behavior without abrupt changes. Cauchy-Riemann equations, a set of partial differential equations involving the real and imaginary parts of both f(z) and its derivative, provide a test for analyticity. If a function satisfies these equations, it guarantees analyticity, opening up a realm of powerful theorems and properties. Understanding analytic functions is key; many of our conformal maps will utilize these functions directly.

The concept of complex differentiation is slightly different than real differentiation. We are now looking at limits in the complex plane. The derivative of f(z) at a point z, denoted f'(z), represents the rate of change of f(z) at that point. The ability to define derivatives extends the concept of calculus into the complex world. Cauchy's integral formula, elegantly linking the value of an analytic function at a point to a contour integral around that point, is a cornerstone theorem. This formula allows us to express the function's value at a point directly in terms of its values on a closed curve surrounding the point. This powerful tool simplifies many calculations and is fundamental to our understanding of complex function behavior.

Among many analytic functions, elementary functions like polynomials, exponential functions, and trigonometric functions have direct analogs in the complex domain. These functions behave similarly to their real counterparts but exhibit unique characteristics in the complex plane. For example, the exponential function in the complex plane maps straight lines to spirals, demonstrating the richer geometric transformations possible. Likewise, trigonometric functions exhibit periodic behavior, leading to their essential role in various applications. These functions aren't merely extended to the complex plane; their complex counterparts reveal hidden properties not readily apparent in the real domain, and their utilization in conformal mappings will greatly simplify geometric problems.

Branch cuts are a crucial element in understanding multi-valued complex functions. Many complex functions, like the complex logarithm or the complex power function, don't have a single unique value for a given input. Branch cuts are lines or curves that separate different branches of these functions, providing a framework to define single-valued branches. Choosing a particular branch cut leads to a well-defined function that is continuous and analytic everywhere within the chosen branch. Understanding the impact of branch cuts on function behavior is essential for avoiding inconsistencies in our conformal mapping processes. We must carefully consider branch cuts when choosing a transformation, ensuring that our map is consistently defined within the region of interest. Misunderstanding branch cuts can lead to unexpected results and incorrect mapping behaviors.

Finally, the concept of conformal mapping lies in the preservation of angles. A conformal transformation preserves angles between curves when mapping from one complex plane to another, reflecting the complex function's analyticity. This angle-preserving characteristic is the very core of our exploration. Many physical phenomena, particularly those governed by Laplace's equation, maintain their properties even under conformal transformations. This principle enables us to solve problems in complex geometries by transforming them to simpler ones where solutions are readily obtained, then transforming the solution back to the original geometry. This concept simplifies solving boundary-value problems that would otherwise be analytically intractable, providing valuable tools for addressing otherwise challenging geometries. The angle-preserving properties are a key advantage to the methods we will use in the following chapters.

1.2 Conformal Transformations

Conformal transformations are a cornerstone of complex analysis, offering powerful tools for solving problems across diverse fields. Their defining characteristic – preserving angles – makes them particularly useful in situations where angular relationships are paramount. Think of fluid flow around an airfoil, where the smooth transition of fluid necessitates preserving angles at the surface. This angle-preserving property, a consequence of the Cauchy-Riemann equations, is what makes conformal mappings so effective in transforming complex geometries into simpler, more analytically tractable ones.

The elegance of conformal maps lies in their ability to simplify complex boundary conditions. Imagine trying to solve Laplace's equation in a region with an irregularly shaped boundary. Directly tackling this problem can be incredibly challenging. However, by applying a conformal transformation, you can map that irregular region onto a simpler one, like a unit disk or a half-plane, where solving Laplace's equation becomes significantly easier. Once the solution is obtained in the simpler domain, the inverse transformation yields the solution in the original, complex geometry.

Several elementary conformal maps provide a foundational toolkit for many applications. Möbius transformations, for instance, are the most general class of conformal maps that take circles and lines to circles and lines, providing flexibility in transforming various shapes. They find application in diverse fields, from mapping geographical regions to analyzing optical systems. Exponential and logarithmic maps are crucial for handling situations involving periodic or logarithmic behaviors, respectively. For example, the logarithmic map transforms wedge-shaped regions into half-planes, a conversion valuable in solving problems involving flow around sharp corners. Similarly, power functions introduce rotational symmetry, providing another method to adjust complex geometries into simplified ones.

The Schwarz-Christoffel transformation represents a particularly powerful tool for mapping polygons to simpler regions. This transformation directly connects the angles of a polygon to the parameters of the mapping function. This direct relationship allows for the precise mapping of highly irregular polygons onto a canonical region, typically the upper half-plane. This method, however, often necessitates numerical methods for its implementation, especially for polygons with many sides or intricate shapes. Techniques such as iterative methods and Newton-Raphson techniques are commonly employed to find optimal parameters and ensure accurate mapping. Even with these numerical approaches, understanding the limitations of the methods and potential errors is crucial. For instance, the convergence of numerical solutions can be dependent on the polygon's shape and the chosen algorithm.

Beyond the basic transformations, understanding the underlying theory is crucial. The concept of analytic functions is paramount, as conformal maps are defined by analytic functions, which are infinitely differentiable and satisfy the Cauchy-Riemann equations. These equations ensure the preservation of angles and the existence of a well-defined derivative throughout the domain. Conversely, any map that violates these equations will not be conformal and would not be applicable for applications relying on angle preservation. Furthermore, the Riemann Mapping Theorem guarantees the existence of a conformal map between any two simply connected domains, provided neither is the entire complex plane, adding a powerful theoretical underpinning to the applicability of these transformations.

The applications of conformal transformations extend far beyond the theoretical realm. In fluid dynamics, for example, they enable the analysis of potential flows around complex geometries like airfoils. By mapping the airfoil's shape to a simpler shape like a circle, analysis of flow patterns becomes significantly easier, allowing for the accurate prediction of lift and drag. Similarly, in electrostatics, conformal transformations simplify the calculation of electric fields and capacitances in systems with intricate geometries. By mapping the geometry onto a simpler domain, calculation of potentials and field strengths is significantly simplified. This can be crucial in designing efficient capacitors or analyzing electric fields in complex microelectronic devices. Their impact spans several other domains, from heat transfer to elasticity and beyond. However, choosing the correct transformation depends greatly on the specifics of the problem and the geometry at hand, requiring careful consideration and understanding of the properties of different mappings.

2 Elementary Conformal Maps

Möbius transformations, arguably the simplest yet profoundly powerful conformal maps, are the cornerstone of many complex analysis applications. Defined by the equation w = (az + b)/(cz + d), where a, b, c, and d are complex constants and ad \item bc 0, they exhibit remarkable properties. They map circles and lines to circles and lines, a characteristic that makes them invaluable for transforming complex geometries into simpler, more manageable forms. Understanding their action on the complex plane is crucial; for example, they can invert points, rotate, translate, and scale, all within a single transformation. This versatility makes them a vital tool for solving boundary value problems, as you'll see later in the book. Mastering their properties is key to unlocking more advanced techniques later on.

The exponential map, w = ez, provides a different perspective, showcasing the mapping of infinite strips to the entire complex plane excluding the origin. This transformation is particularly useful when dealing with periodic functions or problems involving infinite regions. Imagine trying to analyze fluid flow around an infinitely long cylinder; the exponential map can simplify this problem considerably by transforming the infinite region into a more manageable finite one. Consider the transformation of a vertical strip, for instance; the mapping unfolds this confined area onto the complex plane, excluding zero, thus changing the scale of the problem to a far more approachable dimension. Practice visualizing these transformations; it's the best way to internalize their practical implications.

Logarithmic maps, the inverse of exponential maps, perform the opposite transformation, mapping the punctured plane onto an infinite horizontal strip. This provides an essential tool for analyzing problems with logarithmic singularities. These singularities often appear in physical scenarios; for example, a point charge in electrostatics creates a logarithmic potential field. The logarithmic map allows us to transform the vicinity of such singularities, facilitating analysis and problem-solving in an otherwise intractable situation. By transforming this complex space into a more manageable domain we can understand its behavior far more easily. Think of it as unwinding a spiral; the seemingly chaotic arrangement becomes orderly and manageable once mapped.

Power functions, defined by w = zn where n is a positive integer, offer a pathway to create n-sheeted Riemann surfaces and branch points, crucial concepts in more advanced conformal mapping applications. Consider the simple square root function, w = z; this seemingly basic transformation maps the plane into two overlapping sheets. You'll gain deeper insights into branch points and branch cuts when working with power functions. This foundation sets you up for tackling more complex problems involving multi-valued functions and Riemann surfaces which will be discussed in the later chapters. These transformations allow us to view complex functions in a more geometrical way, allowing for a richer understanding of their properties.

Root functions, w = z1/n, are equally significant, as they effectively unravel these n-sheeted Riemann surfaces back into the standard complex plane. This 'unwrapping' is essential for several applications, particularly in problems involving singularities and branch cuts. Understanding how these functions resolve multi-valued functions into a single-valued representation is crucial for the effective manipulation of complex equations. Think of these functions as the inverse of power functions and you'll develop a strong grasp on the interconnectedness and balance between these transformations. Practice manipulating different forms of these equations and you'll quickly master their use. In essence, the mastery of these elementary maps is a springboard to a deeper understanding of more complex conformal mappings.

2.1 Möbius Transformations

Möbius transformations, also known as fractional linear transformations, are fundamental conformal maps possessing remarkable properties. They are defined by the formula w = (az + b) / (cz + d), where a, b, c, and d are complex constants, and ad – bc 0. This seemingly simple expression unlocks a wealth of geometric transformations. Understanding their behavior is crucial for mastering conformal mapping techniques. Their power lies in their ability to map circles and lines to circles and lines—a property that greatly simplifies many problems in complex analysis and its applications.

Consider the special case where c = 0. This yields a simpler transformation, $w = (az + b) / d$, which represents a combination of scaling, rotation, and translation. This highlights how even specific instances of Möbius transformations can represent meaningful geometric operations. Such transformations are easily visualized and understood, forming the basis for understanding more complex variations. These basic transformations lay the groundwork for tackling more intricate mappings.

The general Möbius transformation, however, offers much greater flexibility. It can map any point in the complex plane, except for $z = -d/c$ (the pole), to another unique point in the complex plane. This bijective mapping is a key feature, ensuring that no information is lost during the transformation. This property is essential for solving boundary value problems, as it allows for a consistent mapping between the original domain and the transformed domain. Mapping boundaries accurately is key to solving these kinds of problems.

Furthermore, Möbius transformations possess the remarkable property of composing well. This means that the composition of two Möbius transformations is itself a Möbius transformation. This characteristic simplifies complex sequences of transformations, as the resultant map can always be expressed in the concise form of a single Möbius transformation. The ability to combine transformations efficiently is invaluable in solving complex engineering problems where multiple stages of mapping might be involved.

A powerful application of Möbius transformations lies in the mapping of regions. They can, for instance, transform the upper half-plane to the unit disc. This transformation is particularly useful in solving boundary value problems, as the unit disc often leads to simpler solutions than the upper half-plane. For example, the Dirichlet problem, which involves specifying boundary conditions on a region, can become significantly easier to solve after a suitable transformation to the unit disc.

Let's consider an example. Suppose we want to map the upper half-plane to the interior of the unit disc. This specific mapping has wide applications, from simplifying certain types of differential equations to aiding in complex simulations. The particular Möbius transformation that achieves this mapping is given by a specific formula derived from the general form. Choosing the correct parameters (a, b, c, and d) within the equation is crucial, but the fundamental principle stays the same.

Another noteworthy application involves the mapping of polygons. While the Schwarz-Christoffel transformation is more directly applicable to mapping polygons, Möbius transformations are frequently used as preliminary transformations to simplify the overall mapping process. This preprocessing step significantly reduces the complexity of the subsequent Schwarz-Christoffel transformation.

In summary, Möbius transformations are not merely abstract mathematical constructs; they are powerful tools with significant practical implications. Their ability to map circles and lines to circles and lines, coupled with their composition property and their role in simplifying boundary value problems, makes them indispensable in various branches of engineering and physics, specifically within fluid dynamics, electrostatics, and heat transfer analysis. Mastering these transformations is crucial for any serious student of complex analysis. Their use greatly simplifies complex problems, offering elegant solutions where direct methods often fall short. Understanding their versatility is paramount for effective problem-solving.

2.2 Exponential & Logarithmic Maps

The exponential map, w = ez, transforms the complex plane in fascinating ways. Imagine a horizontal strip in the z-plane, say 0 Im(z) < 2. This seemingly simple strip, when mapped by the exponential function, explodes into the entire complex w-plane, excluding the origin. Each horizontal line within the strip maps to a ray emanating from the origin, while vertical lines map to circles centered at the origin. This illustrates the expansive nature of the exponential map, transforming a bounded region into an unbounded one. Understanding this mapping is crucial for solving boundary value problems involving infinite regions, a common scenario in many engineering applications. The periodic nature, with its repetitive mapping of strips, underscores the importance of carefully choosing the domain of interest when utilizing this transformation.

Logarithmic mappings, the inverse of exponential maps, offer a powerful tool for tackling problems involving regions with angular boundaries. The principal branch of the logarithm, w = Log(z), for instance, maps the punctured plane (excluding the origin and the negative real axis) onto the horizontal strip - < Im(w) . Notice the reversal: the unbounded region becomes a bounded one. The intricate details of branch cuts, inevitable with multi-valued functions, must be meticulously handled. Misunderstanding these cuts can lead to incorrect mapping results, and thereby, inaccurate solutions to the problem at hand. In the process, you are effectively unwinding what was essentially a logarithmic spiral. This seemingly simple inverse transformation significantly simplifies the analysis of regions with sharp corners and infinite boundaries.

Consider a specific scenario, calculating the electric field around a charged conducting cylinder. The cylindrical geometry presents difficulties if we solve it directly in its natural Cartesian coordinates. However, by applying the logarithmic map, we can transform the problem into a simpler geometry where solving Laplace's equation, governing the electric potential, becomes dramatically easier. The transformation maps the exterior of the cylinder to a strip in the w-plane, and suddenly, solutions that were otherwise impossible become accessible through elementary methods. This simplification is a testament to the power of conformal mappings in solving seemingly intractable problems, transforming complex geometries into easily manageable ones. The elegance of this approach makes it an invaluable tool in electrostatics and other fields.

Let's look at another application, fluid flow around an airfoil. The complex shape of the airfoil makes the analysis of the flow extremely challenging. Using conformal mappings, particularly the Joukowski transformation (which incorporates both exponential and other transformations), we can transform the airfoil into a simpler shape, such as a circle. Analyzing the flow around the circle is significantly easier, and we can then reverse the transformation to obtain the flow pattern around the original airfoil. This allows us to design more efficient and effective airfoils using this technique. This method forms the cornerstone of much of aerofoil design and analysis, highlighting the practicality of theoretical techniques like these conformal mappings.

The power of exponential and logarithmic maps extends beyond mere geometric transformations. They provide a crucial link between different mathematical descriptions of physical phenomena. For example, in heat transfer, the temperature distribution in certain geometries can be more readily solved using logarithmic transformations. Similarly, in wave propagation, certain wave patterns can be analyzed more effectively by understanding how these maps alter the boundaries and, consequently, the wave behavior itself. It's this transformative power, altering the mathematical landscape, that makes these conformal mappings so valuable. The potential applications extend even further, showcasing the inherent utility in mathematics and its applications to a wide range of real world problems. The possibilities are as expansive as the complex plane itself. Finally, remember that while these mappings offer remarkable simplification, numerical methods are often required for practical implementation, particularly when dealing with complex geometries or boundary conditions. Software packages are readily available that can assist in implementing these transformations, enabling the effective application of this rich mathematical framework. The ease with which you can now solve previously difficult problems underscores the value of understanding conformal mappings and choosing appropriate transformations for each application. It is in this intersection of theoretical elegance and practical application that the real power of these maps truly shines through.

2.3 Power & Root Functions

Power functions, represented as w = z where 'n' is a complex constant, offer a straightforward yet potent method for conformal mapping. The effect on the complex plane depends critically on the value of 'n'. For instance, when 'n' is a positive integer, the transformation maps the unit circle onto itself, but with 'n' rotations around the origin. Consider n=2; the upper half-plane is mapped onto the entire plane, excluding the negative real axis. This doubling effect is easily visualized. It's crucial to understand that the mapping is not one-to-one for integer n greater than 1; multiple points in the z-plane map to the same point in the w-plane. This multi-valued nature requires careful consideration when solving boundary value problems. For instance, if you are mapping a region with a specific boundary condition you may need to account for this multi-valued behavior by carefully selecting the appropriate branch of the function. Practical applications often necessitate careful examination of the branch cuts to prevent ambiguity.

Root functions, conversely, present a more intricate transformation. They are the inverse of power functions, with $w = z\hat{}(1/n)$, again, with 'n' a complex constant. Unlike their power function counterparts, root functions are multi-valued for all integer n > 1. This multi-valued nature introduces a considerable challenge in applications. Each value of n generates 'n' distinct branches, each mapping different sectors of the z-plane onto the entire w-plane. The selection of a specific branch is vital; otherwise, any interpretation of the results remains ambiguous, and ultimately meaningless. One method for dealing with this inherent ambiguity is to restrict the domain, effectively selecting only one branch, thereby transforming a multi-valued function into a single-valued function. In this manner, one can still utilize the transformation for specific applications, provided such restrictions do not compromise the accuracy and relevance of the result.

The effect of these transformations on simple regions, such as sectors and half-planes, is readily understandable, and these serve as excellent starting points for gaining intuition and a strong conceptual understanding. For example, consider mapping a sector of the z-plane using a power function. The result will be a sector with an angle that is a multiple of the original angle, directly dependent on the value of n. For root functions, the reverse occurs; a sector in the z-plane will be mapped to a larger sector in the w-plane. This simple yet effective visualization enables one to grasp the essence of the transformations involved, laying the groundwork for tackling more challenging problems. It's beneficial to practice visualizing these mappings through graphical representations, particularly with the aid of software that can render complex mappings.

To effectively use power and root functions in conformal mapping, a thorough understanding of branch cuts is necessary. Branch cuts are lines or curves in the complex plane where the function becomes discontinuous; across these cuts, the function's values abruptly change. Selecting the appropriate branch cut is fundamental for ensuring consistent and meaningful results. It allows you to define the particular branch you are working with, limiting ambiguity. Incorrect branch cut selection may lead to completely erroneous results, especially when dealing with regions that cross the cut. It's therefore crucial to carefully analyze the problem's geometry and the function's behavior to determine the most suitable branch cut. This thoughtful approach prevents misinterpretations of solutions, promoting accurate results and furthering a precise comprehension.

In practical applications, the choice between power and root functions often depends on the specific problem. If you are seeking to map a simpler region to a more complex one, power functions may be better suited. If the objective is to transform a complex area into a simpler one for easier analysis, root functions may be preferable. The multi-valued nature of root functions often necessitates the introduction of branch cuts to define a unique solution within a given domain. Such considerations highlight the critical role of meticulous analysis and careful selection of both functions and branch cuts to acquire accurate and reliable solutions. Consider a classic example: mapping a semi-infinite strip to a half-plane. A clever combination of power and logarithmic functions can achieve this transformation effectively. This exemplifies how different elementary conformal mappings can be combined to address more complex problems in boundary value contexts. This approach highlights the versatility and strength of conformal mappings in addressing a wide array of geometrical situations. The application of these transformations to more complex geometries often involves finding a suitable coordinate system and applying a combination of transformations and boundary conditions. It is often necessary to iterate and adjust approaches until a satisfactory solution is reached.

Ultimately, mastering power and root functions is paramount to a proficient understanding and application of conformal mappings. These functions serve as building blocks for tackling more complex transformations and provide a foundation for understanding the underlying principles of conformal mapping. They are simple yet powerful tools that help tackle difficult problems in physics and engineering. It is through repeated application and careful consideration of the mathematical intricacies that these functions reveal their full potential in solving real-world problems. This foundational knowledge provides the necessary groundwork for advancing into more sophisticated techniques, significantly enhancing your comprehension and skill in conformal mapping.

3 Schwarz-Christoffel Transformations

The Schwarz-Christoffel transformation stands as a powerful tool within the arsenal of conformal mapping techniques. Unlike the simpler transformations explored previously, this method allows us to map the interior of a polygon in the complex plane onto the upper half-plane. This seemingly simple transformation holds immense practical value, allowing the solution of otherwise intractable boundary value problems in diverse fields. Its application hinges on the understanding that the derivative of the transformation is expressed as a product of terms involving the prevertices – points in the upper half-plane that map to the vertices of the polygon. This formula, elegant in its simplicity, belies the often considerable computational effort required for its implementation.

Determining the precise prevertices remains a crucial challenge. For simple polygons, the calculation may be straightforward, offering a direct analytical solution. However, for more complex polygons with numerous sides, finding these prevertices often necessitates numerical methods, demanding a careful iterative approach to minimize error. This process commonly involves sophisticated algorithms that minimize a cost function–the difference between the desired polygon shape and that produced by a given set of prevertices. It is important to stress that the accuracy of the mapping directly depends on the precision achieved in determining these prevertices. Improper determination leads to inaccurate mapping and, consequently, unreliable solutions to the underlying physical problems.

Numerical methods, while powerful, introduce inherent limitations. Computational time often increases significantly with the complexity of the polygon, potentially demanding substantial processing power. Furthermore, numerical instability can occur, particularly when dealing with polygons exhibiting extreme aspect ratios or very close-to-coincident vertices. These situations require careful attention and possibly the utilization of advanced techniques. Such techniques might involve adaptive mesh refinement or the application of regularization procedures to stabilize the computations. A thorough understanding of these numerical considerations is crucial for ensuring reliable and accurate results when applying the Schwarz-Christoffel transformation.

Beyond the purely numerical aspects, the selection of appropriate parameters is paramount. These parameters often include the choice of the prevertices and the scaling factor, which influences the size and orientation of the mapped polygon. Careful consideration is required to optimize these choices, often balancing computational efficiency with the desired accuracy. Improper choices may lead to either computationally expensive calculations or inaccurate representations of the desired polygon. Sensitivity analysis plays a pivotal role here, allowing for assessment of the impact of these parameter choices on the overall accuracy of the transformation. This systematic evaluation is key to ensuring robustness and reliability in the mapping procedure.

Finally, practical applications demand a realistic appreciation of limitations. While incredibly powerful, the Schwarz-Christoffel transformation isn't a panacea. The method is primarily suited for polygons; mapping more complex regions requires more advanced conformal mapping techniques or alternative approaches entirely. Moreover, the accuracy of the resulting mapping is inherently sensitive to the precision with which the prevertices are determined. Therefore, it's essential to employ robust numerical methods and rigorously test the accuracy of any resulting solutions. Remember, the Schwarz-Christoffel transformation provides a powerful tool, but skillful application and a comprehensive understanding of its limitations are vital for reliable and successful results. Understanding these subtleties ensures the responsible and effective use of this potent mathematical technique in solving complex real-world problems.

3.1 Mapping Polygons

The Schwarz-Christoffel transformation provides a remarkably powerful tool for mapping polygons onto the upper half-plane. Its strength lies in its ability to handle complex geometries, transforming intricate shapes into simpler domains where calculations become significantly more manageable. This transformation is particularly valuable in solving boundary value problems within these irregularly shaped regions, problems otherwise intractable using more elementary techniques. The essence of the method hinges on a carefully constructed integral formula that directly relates the pre-image points in the polygon to their corresponding points in the upper half-plane. Understanding this connection is pivotal to uti-

lizing the Schwarz-Christoffel transformation effectively.
This integral relies on the specific pre-image vertices of the polygon and their associated interior angles. Each vertex contributes a crucial factor to the integral, directly reflecting its influence on the overall mapping. Finding the correct correspondence between vertices and their images is therefore a key step, often requiring careful geometric analysis and intuition. It's not simply a mechanical process; solving real-world problems requires a combination of mathematical rigor and insightful interpretation. This iterative process allows refinement and closer approximation to the desired mapping.
Consider, for instance, mapping a rectangle onto the upper half-plane. This seemingly straightforward task highlights several crucial aspects of the transformation. Initially, we identify the vertices of the rectangle, and these serve as input points for the Schwarz-Christoffel formula. We then select their corresponding points in the upper half-plane. This choice isn't arbitrary; it significantly impacts the resulting map and the relative scaling and positioning of the mapped rectangle. The selection process requires careful consideration of the relative positions and the desired orientation. A slight adjustment in the mapping points can dramatically alter the final mapped shape.
The pre-vertex angles of the polygon play a critical role. Each interior angle directly influences the integrand, affecting the mapping's overall form. An acute angle will cause a corresponding compression in the mapped shape, while an obtuse angle will lead to stretching or expansion. Understanding how these angles translate into the final map is crucial for constructing the accurate transformation. This means a deep understanding of geometry and how it interacts with complex analysis is

key to success.

Numerical methods are often indispensable when dealing with the Schwarz-Christoffel transformation. Exact solutions are rare, especially for polygons with more than a few sides. Numerical techniques such as iterative algorithms provide effective approximations, allowing us to work with complex polygons where analytical solutions are impractical. These approximations provide a powerful tool for practical applications. Software packages are often utilized for these calculations, significantly simplifying the process and reducing the computational workload. The choice of the numerical method hinges on the problem's specific needs and desired accuracy.

Implementing the Schwarz-Christoffel transformation involves more than just the application of formulas. Practical considerations significantly impact the accuracy and efficiency of the process. Careful selection of parameters, such as the location of the pre-image points in the upper half-plane, can dramatically affect the result. Numerical stability is another significant consideration, particularly when dealing with ill-conditioned problems or polygons with extreme aspect ratios. In such cases, a robust numerical scheme is essential for avoiding inaccurate or unstable results. Careful attention to these practical aspects is vital for ensuring the successful application of the transformation.

The power of the Schwarz-Christoffel transformation extends far beyond theoretical exercises. It finds widespread application in various engineering disciplines. For instance, in fluid dynamics, it simplifies the analysis of fluid flow around complex shapes. It allows for precise analysis of complex flow patterns around airfoils, streamlining the design process for aerodynamic efficiency. In electrostatics, this transformation aids in accurately determining electric field distributions around irregularly shaped conductors, essential for designing efficient and reliable electrical components. Its significance in diverse fields highlights its versatility and its power as a problem-solving tool. This demonstrates its practical utility and the potential for further innovation based on this fundamental concept.

3.2 Applications & Examples

Consider a long, rectangular channel with a sudden constriction. Analyzing fluid flow through this geometry using traditional methods is incredibly complex. However, using a conformal mapping, specifically the Schwarz-Christoffel transformation, we can simplify the problem dramatically. By mapping the channel onto the upper half-plane, the complex flow equations transform into a much more manageable form, allowing for easier calculation of pressure, velocity, and other key flow parameters. This is a direct application of the theoretical principles we've covered, demonstrating the power of conformal mappings in practical engineering scenarios. The transformation's effectiveness hinges on accurately defining the polygon's vertices, each representing a critical point in the channel's geometry.

The efficiency of electrical transmission lines benefits significantly from optimized designs. One common challenge is designing lines that minimize energy loss over long distances. The intricate shapes involved, often influenced by terrain and infrastructure constraints, can make direct analysis of the electric field incredibly challenging. Here, conformal mappings, such as the Joukowski transformation, provide a powerful tool.

This transformation allows for the mapping of a complex cross-section, representing the transmission line, onto a simpler shape, such as a circle, where calculations become significantly easier. After solving the problem on the simplified geometry, the solution is then mapped back to the original, complex geometry, providing critical data for effective line design. This method is used extensively in the optimization of high-voltage transmission lines, ensuring efficient energy distribution.

Let's delve into the fascinating realm of aerodynamics. Designing efficient airfoil profiles requires understanding complex airflow patterns. Traditional methods can be computationally expensive and often lack the accuracy needed for sophisticated designs. Conformal mapping offers a more elegant solution. The Joukowski transformation, for instance, allows for the generation of airfoil shapes from simple circles. By transforming the flow around the circle (a problem with a well-known analytical solution), we obtain a solution for the flow around the complex airfoil shape, providing valuable insights into lift, drag, and other critical aerodynamic characteristics. This approach is incredibly powerful, allowing engineers to create high-performance airfoils with minimal computational effort, significantly reducing design and testing times. Moreover, understanding the influence of specific geometric parameters on aerodynamic performance becomes much simpler.

Imagine a capacitor with unusual electrode shapes. Calculating its capacitance accurately through conventional methods would prove extremely challenging, requiring complex numerical integrations. However, the power of conformal mapping shines through once more. By employing appropriate transformations, we can map the irregular capacitor geometry into a simpler shape, such as two parallel plates. This simplification drastically simplifies the computation of the capacitance. The capacitance of the parallel-plate capacitor is well-known, facilitating a straightforward calculation. Then, by applying the appropriate scaling factors derived from the mapping itself, the capacitance of the originally complex capacitor can be efficiently determined, offering a precise result without resorting to computationally demanding methods. This is a real-world example where the elegance of mathematical transformations leads to practical engineering solutions.

The design of efficient heat exchangers relies heavily on understanding heat transfer across complex geometries. Traditional methods often struggle with the intricate flow patterns and temperature distributions within such systems. Here, conformal mappings offer a powerful tool for optimizing the design of heat exchangers. By mapping the complex heat exchanger geometry onto a simpler geometry, we can simplify the governing equations, enabling more efficient solutions and predictive analysis. This simplification significantly reduces computational time and allows for a clearer understanding of the influence of geometric parameters on heat transfer performance. This leads to the design of highly efficient heat exchangers, crucial in various applications from power generation to chemical processing. The transformation employed depends heavily on the specific geometry but common choices often include variations of the Schwarz-Christoffel transformation for more irregular shapes, or even simpler transformations if the geometry allows for it.

The applications of conformal mapping extend beyond these examples. In groundwater flow modelling, for instance, conformal mapping can simplify the solution of the Laplace equation for complex aquifer geometries. Similarly, in the design of electromagnetic devices, such transformations can simplify calculations of magnetic fields. The versatility and efficiency of these methods make them invaluable tools for a wide range of engineering and scientific problems. This highlights the broad applicability of conformal mapping across diverse disciplines and the significant advantages it provides over traditional techniques. Understanding these fundamental mappings is essential for anyone seeking to solve complex problems within these fields. Mastering this technique is not just about mathematical dexterity but about unlocking efficient solutions to real-world challenges.

3.3 Numerical Methods

Finding the Schwarz-Christoffel transformation's parameters often requires numerical methods. The transformation itself, elegantly mapping polygons onto the upper half-plane, hinges on the precise determination of these parameters – prevertices and their corresponding angles. Analytical solutions are rare; numerical approaches are the norm. Consider a polygon with numerous sides: the complexity of an analytical solution escalates exponentially. Numerical techniques offer a practical path to solve this challenging problem.

Iterative methods prove particularly useful. The Newton-Raphson method, for example, refines an initial guess for the parameters through repeated iterations, converging towards a solution that minimizes a chosen error function. This function typically quantifies the discrepancy between the desired polygon and the image produced by a given parameter set. The method's effectiveness depends on a shrewd initial guess; a poor starting point might lead to slow convergence or divergence altogether. Sophisticated algorithms exist that generate such good initial guesses.

Another powerful technique employs a boundary integral equation method. This method transforms the problem into a system of linear equations, solvable using established numerical linear algebra techniques. This approach handles polygons with complicated shapes and an arbitrary number of vertices effectively. The elegance lies in its ability to reduce a complex geometric problem to a solvable algebraic one. The accuracy of the solution is directly related to the discretization of the boundary; finer discretization improves accuracy but increases computational demand.

For those new to numerical analysis, understanding the concepts of convergence and stability is crucial. Convergence refers to how closely the iterative method's results approach the true solution. A method converges slowly when many iterations are needed for acceptable accuracy; it diverges if it fails to converge to a solution at all. Stability is related to the method's resilience to small errors – both in the initial guess and within the calculation process itself. Unstable methods amplify errors, potentially leading to inaccurate or nonsensical results. Choosing a stable and efficient method is paramount for reliable results.

Software packages offer pre-built routines that implement several numerical methods for the Schwarz-Christoffel transformation. These packages often include error handling and advanced convergence acceleration techniques, making the process more efficient and robust. The choice of software depends on the specific problem and available resources; some are more user-friendly than others. Familiarity with at least one such package is invaluable for practical applications.

Beyond the direct computation of parameters, numerical methods also play a role in evaluating the transformation itself. The Schwarz-Christoffel integral, central to the transformation, is often impossible to solve analytically. Numerical integration techniques, like Gaussian quadrature or trapezoidal rules, are needed to approximate its value for specific points within the mapped region. The choice of integration method influences the accuracy and speed of the computation. An accurate evaluation of the integral is essential for plotting the transformed region and extracting useful information.

The accuracy of the numerical method significantly impacts the final results. While the theoretical framework of the Schwarz-Christoffel transformation is elegant, the practical implementation necessitates numerical approximations. These approximations introduce errors, which need careful management. Understanding the sources of error–such as numerical integration errors, truncation errors in iterative methods, and rounding errors in computer arithmetic–is essential for assessing the reliability of the results. Detailed error analysis is a critical part of any proper application.

Remember, numerical solutions are approximations. Although refined methods minimize errors, a degree of uncertainty always remains. Presenting results, therefore, requires transparency. Clearly stating the numerical method used, its parameters, and an estimate of the solution's error are crucial for responsible usage and interpretation of the numerical results. This transparency allows others to assess the reliability and limitations of the findings. This is vital within the larger framework of applied mathematics.

3.4 Advanced Techniques

The Schwarz-Christoffel transformation, while powerful, presents challenges. Its integral representation often lacks closed-form solutions, necessitating numerical methods. For instance, consider mapping a polygon with many sides; the integrand becomes increasingly complex, demanding sophisticated numerical integration techniques. Gaussian quadrature, for example, proves effective for smooth integrands, but for highly oscillatory integrands encountered in irregular polygons, adaptive quadrature methods, which adjust the integration points based on the integrand's behavior, offer greater accuracy and efficiency. These methods dynamically refine the integration grid in regions of rapid change, ensuring precise evaluation of the Schwarz-Christoffel integral even for complex polygon shapes.

However, numerical integration is only one hurdle. Determining the pre-vertices, the points in the parameter plane that map to the vertices of the polygon, often requires iterative procedures. Newton-Raphson methods, commonly employed, demand good initial guesses to ensure convergence to the correct solutions. Poor initial guesses may lead to divergence or convergence to an incorrect solution. Techniques such as the method of successive approximations, while slower, provide more robust convergence properties. Combining these with a careful analysis of the polygon's geometry can often improve the chances of finding the correct pre-vertices. Remember, accuracy here directly impacts the accuracy of the final conformal map.

Beyond numerical integration and pre-vertex determination, consider the handling of degenerate polygons. These polygons may possess collinear vertices or angles exceeding radians, presenting mathematical singularities in the Schwarz-Christoffel formula. Special techniques must be implemented to handle such cases, often involving modifications to the integration path or reformulation of the problem using limiting arguments. Careful consideration of these degenerate cases is paramount for the successful application of the Schwarz-Christoffel transformation. Failure to account for these subtleties can lead to erroneous results and inaccurate mappings.

Furthermore, practical implementation often involves dealing with mapping functions that are not just numerically challenging but also analytically complicated. Sometimes the resulting conformal mapping is implicitly defined; this means that expressing the mapping function explicitly may be intractable, leading to reliance on iterative solution methods for finding the image points. This can add another layer of complexity and computational burden to the overall process. Therefore, understanding and optimizing these iterative procedures is crucial. Techniques for accelerating convergence, such as Aitken's delta-squared process or Steffensen's method, can significantly improve the efficiency of these computations, which is critical when dealing with intricate geometries or high-precision requirements.

Moreover, error analysis is essential in any numerical implementation. The inherent errors in numerical integration and the iterative determination of pre-vertices propagate throughout the mapping process. Rigorous error estimations must be performed to quantify these inaccuracies. Understanding the sources of error – truncation errors from numerical integration, rounding errors from finite-precision arithmetic, and convergence errors from iterative solvers – allows you to assess the reliability and accuracy of the obtained conformal map. Adaptive methods for both integration and iteration, along with careful monitoring of error bounds, are necessary for generating high-quality, reliable mappings. Remember, a robust error analysis is not simply a theoretical exercise; it's a vital step in ensuring the practical utility of the Schwarz-Christoffel transformation.

Finally, remember that even with careful implementation, limitations exist. The Schwarz-Christoffel mapping, while powerful, is limited to mapping polygons. More complex geometries require different approaches. Also, the mapping might introduce significant distortion in certain regions of the plane, potentially obscuring fine details or exaggerating less important features. This necessitates a thorough understanding of the limitations of the mapping and the selection of an appropriate technique suited to the problem at hand, and careful consideration of the trade-offs between accuracy and computational cost. Often, different methods may be combined, utilizing the strengths of several techniques to overcome individual weaknesses, resulting in a more robust and efficient solution. The successful application of these advanced techniques hinges on a deep understanding of the underlying mathematics and a careful assessment of the specific challenges presented by the problem.

3.5 Practical Considerations

The Schwarz-Christoffel transformation, while powerful, presents several practical challenges. Accurate computation of the pre-vertices, those points in the upper half-plane that map to the vertices of the polygon, often requires iterative numerical methods. These methods can be computationally expensive, particularly for polygons with many vertices or those exhibiting significant geometric complexity. Careful consideration of numerical precision is crucial, as errors in the pre-vertex calculations can propagate and significantly distort the resulting mapping.

For instance, consider mapping a polygon with a large number of sides. Determining the precise locations of the pre-vertices necessitates solving a system of nonlinear equations. While algorithms exist to handle this, the computational burden can become substantial, especially when high accuracy is required. Furthermore, ill-conditioned polygons—those where vertices are clustered closely together—can further exacerbate numerical instability, leading to inaccurate or unreliable mappings. Understanding these limitations is essential for successful application of the method.

Software packages that implement the Schwarz-Christoffel transformation often incorporate sophisticated algorithms to mitigate these challenges. However, users should be aware that the accuracy of the resulting conformal map depends critically on the quality of the input data and the choice of numerical methods. It's often advisable to conduct convergence studies and sensitivity analyses to gauge the reliability of the obtained results. These are crucial steps not to be overlooked during the application phase.

Beyond computational challenges, practical applications often involve additional considerations. For example, real-world geometries are rarely perfectly represented by polygons. Approximating a complex geometry with a polygon may introduce errors into the resulting conformal mapping. The accuracy of this approximation directly impacts the utility and reliability of the resulting analysis. A careful assessment of the trade-off between accuracy and computational cost is often necessary when choos-

ing an appropriate polygon representation.
Moreover, the Schwarz-Christoffel transformation is typically defined only for simply connected polygons. In cases dealing with multiply connected regions, the direct application of the standard transformation becomes infeasible. Advanced techniques such as the use of multiply connected regions or the application of the method to different components of a complex shape, followed by a stitching process, then become necessary. This involves sophisticated mathematical techniques to overcome the limitations of standard implementations.
This complexity underscores the importance of careful planning and validation when using the Schwarz-Christoffel transformation. A thoughtful approach to polygon selection and numerical implementation can dramatically improve the accuracy and reliability of the method. In cases of highly complex shapes or strict accuracy requirements, alternative mapping techniques or advanced numerical methods may need to be explored. This often necessitates consultation with numerical experts and involves fine-tuning of algorithms for optimal performance. Understanding these nuances separates successful applications from flawed ones.
Therefore, successful implementation demands a nuanced understanding of the inherent challenges and limitations. Prioritize efficient numerical algorithms, meticulously assess the accuracy of the obtained results and, critically, conduct rigorous validation. In complex scenarios, explore alternative approaches to guarantee accurate and dependable solutions. These choices will depend on the project requirements; a high degree of accuracy may necessitate advanced techniques or specialized software packages, whilst simpler projects may only require basic implementations. Remember, despite its elegance and power, the Schwarz-Christoffel trans-

formation is a tool requiring careful handling. A thorough understanding of its theoretical foundations, combined with a pragmatic approach to numerical implementation, is vital for obtaining reliable and meaningful results in various engineering and physics applications. Success hinges upon a comprehensive understanding of its capabilities and limitations, along with attention to detail throughout the entire process of analysis. This detailed knowledge, however, will often prove invaluable in the design and modeling of many complex systems and in the analysis of numerous physical phenomena. This ensures the practicality and utility of the method for a wide range of users and applications, ultimately highlighting its continuing relevance in modern engineering and physics.

4 Boundary Value Problems

The Dirichlet problem, a cornerstone of potential theory, presents us with a region in the complex plane and its boundary values. Our task is to find a harmonic function within this region that matches these prescribed boundary values. Conformal mapping offers an elegant solution strategy. By mapping the complex region onto a simpler one, often a disk or a half-plane, we drastically simplify the problem. This transformation doesn't alter the harmonic nature of the solution; it merely changes its representation. Solving the problem in the simpler domain is significantly easier, and then, using the inverse mapping, we obtain the solution in the original, more complicated domain. Consider, for instance, a problem posed on an irregularly shaped plate. Transforming this plate into a unit circle via a clever conformal map simplifies the calculation exponentially. The resulting solution, mapped back to the original shape, provides the accurate temperature or potential distribution across the plate. The precision achieved by this method is remarkable, demonstrating the power of conformal mapping in solving complex boundary value problems.

We can extend this approach to numerous scenarios, handling a wide range of boundary conditions. Imagine designing an efficient heat sink. The heat dissipation across the fin's surface can be precisely modeled using this technique. The conformal mapping converts the irregularly shaped fin into a more manageable geometry, where the heat equation becomes easier to solve. The accuracy of this method, compared to numerical approximations, is often superior, producing results that better reflect the physical reality. The process offers a profound understanding of the heat distribution, helping engineers optimize the heat sink's design for maximum efficiency. The method is crucial in countless real-world engineering applications.

Moving to the Neumann problem, we deal with a different boundary condition – the normal derivative is specified. Instead of fixing values at the boundary, we prescribe the flux. This scenario is paramount in fluid mechanics, modeling the flow of an ideal fluid around an obstacle. The normal derivative represents the velocity component perpendicular to the boundary; defining this component along the boundary describes the fluid's behavior. Once again, conformal mapping shines. We can map the complicated flow domain, with its complex boundary, onto a simpler domain, where the solution becomes more tractable. The same principle holds – the harmonic nature of the solution remains unchanged, allowing for a straightforward solution in the simpler domain, subsequently transformed back into the original domain.

For example, consider the flow around an airfoil. The complex shape of the airfoil makes direct analytical solution difficult. However, via a conformal mapping, we can transform the airfoil into a circle or a slit. The flow around this simpler geometry is easily described. By mapping this simpler flow back to the original airfoil geometry, we obtain a detailed understanding of the velocity distribution and pressure around the airfoil. This is pivotal in understanding lift and drag. The precision of this method is exceptional, providing engineers with critical information for optimizing airfoil design and performance. This method is widely used in computational fluid dynamics, providing accurate simulations that inform design choices. Think of designing airplane wings – it's impossible to overestimate the importance of such detailed and accurate calculations.

These boundary value problems highlight the immense practical utility of conformal mapping. It's not just a theoretical tool; it's a powerful technique with direct applications in engineering and physics. The transformation's ability to simplify complex geometries while preserving essential properties is at the heart of its success. By applying conformal mappings to diverse problems, engineers gain the ability to solve previously intractable equations, leading to more efficient designs and more accurate simulations. These tools prove their worth by consistently enhancing problem-solving capabilities in various fields. Remember, while the mathematical underpinnings might appear complex, the core idea–simplifying the problem, solving it, and mapping the solution back to the original domain–is surprisingly straightforward. The results, however, are anything but simple: a profound understanding of complex systems.

The choice of the appropriate conformal map depends heavily on the specific problem's geometry. There's no one-size-fits-all solution, demanding careful consideration of the region's shape and the boundary conditions involved. For simple geometries, elementary maps like Möbius transformations often suffice. For more complex geometries, particularly those with polygonal boundaries, the Schwarz-Christoffel transformation, while more computationally intensive, provides a precise mapping. The selection process is an integral part of successfully applying conformal map-

ping techniques. Mastering this selection is crucial for efficient problem-solving, enabling the user to transform the complex problem into a solvable one, paving the way for accurate solutions. The process often involves intuition and experience as much as precise calculation.
Mastering boundary value problems using conformal mapping techniques requires a balance of theoretical understanding and practical application. While the underlying mathematical theory is essential, the true power of these methods lies in their ability to simplify real-world problems. By diligently studying the various conformal maps, their properties, and their applications, we become proficient problem solvers. Practice is key; attempting numerous examples, working through varied geometries, and experimenting with different maps will build your skills and confidence. Understanding the limitations of this technique, like handling certain singularities and selecting the right transformation, is equally significant. It's only by practicing and experimenting that you will really understand the usefulness of this approach, transforming your mathematical knowledge into a valuable problem-solving tool. The more you practice, the more naturally you will choose the right approach.

4.1 Dirichlet Problem

The Dirichlet problem, a cornerstone of potential theory, concerns finding a harmonic function within a given region that matches specified values on its boundary. Imagine a flat, perfectly conductive metal sheet with a specific voltage prescribed along its edges. Determining the voltage at every point within the sheet is precisely the essence of the Dirichlet problem. This seemingly simple scenario has profound implications across various fields, from electrostatics and fluid dynamics to heat transfer and even image processing.

Unlike many mathematical problems, the Dirichlet problem often defies straightforward algebraic solutions, particularly for regions with complex shapes. This is where the power of conformal mappings shines. By employing a suitable conformal transformation, we can map the original, intricate domain onto a simpler, more manageable one, often a unit disk or a half-plane. The transformed problem, while equivalent, becomes significantly easier to solve. This is because harmonic functions remain harmonic under conformal mappings, a truly remarkable property.

Solving the transformed problem usually involves employing techniques such as separation of variables or Fourier series. These established methods are well-suited to the simpler geometries, providing an elegant and efficient path to the solution. Once we've obtained the solution in the transformed domain, the inverse conformal mapping then provides the solution to the original problem. This elegant two-step process dramatically simplifies what would otherwise be an intractable problem. Consider a problem involving an oddly-shaped capacitor; using conformal mapping, you could map its complicated shape to a simpler one, solve the potential there, and map it back, determining capacitance more easily.

The choice of conformal mapping is crucial. Different mappings are suited for different geometries. For instance, Möbius transformations are ideal for handling circular regions, while the Schwarz-Christoffel transformation is uniquely suited for polygonal domains. A careful consideration of the problem's geometry and boundary conditions is paramount in selecting the most appropriate transformation. The skill lies in identifying a conformal map that simplifies the problem significantly, while maintaining mathematical rigor. Improper selection can lead to complicated calculations even more challenging than the original problem. Consider the potential applications in the analysis of heat flow in irregular components. Choosing the correct transformation streamlines the process of obtaining a relevant solution.

Furthermore, numerical methods often play a vital role, especially when dealing with highly irregular boundaries or complicated transformations. Techniques like finite difference or finite element methods can approximate the solution, bridging the gap where analytical solutions prove elusive. These computational methods, however, must be implemented carefully, ensuring that the discretization aligns with the complexities of the conformal mapping and the boundary conditions. Using an iterative numerical solver, you can iteratively approximate a solution for complex geometries, using known mathematical conditions to check accuracy.

The impact of the Dirichlet problem extends far beyond theoretical exercises. Its applications are wide-ranging and critical in diverse engineering disciplines. In electrostatics, it models electric potential in complex geometries, enabling the design of high-performance capacitors and other electronic components. In fluid dynamics, it helps model steady, irrotational flows around objects with complex shapes, informing aerodynamic designs and enhancing predictions of flow behaviors.

The problem isn't confined to these areas. In heat transfer, the Dirichlet problem governs the distribution of temperature within solids with prescribed boundary temperatures, finding applications in the design of heat sinks and thermal management systems. Even in image processing and computer graphics, conformal mapping is leveraged to deform images while maintaining consistency in details, enhancing distortion and

filtering techniques. Consider designing a more efficient airplane wing shape. By utilizing this method, one could model the airflow around the wing with superior accuracy, resulting in a more efficient design.
Mastering the Dirichlet problem through conformal mapping equips you with a powerful toolset for solving complex boundary value problems across a multitude of disciplines. While the initial concepts might seem daunting, a systematic approach–understanding the problem's geometry, choosing an appropriate conformal mapping, solving the transformed problem, and inverting the mapping–leads to elegant and efficient solutions, transforming complex problems into more accessible ones. It's a testament to the power and beauty of mathematical techniques to address complex scenarios, offering practical solutions in various fields. Ultimately, understanding the method allows for the efficient solutions of a variety of problems, from the simple to the profoundly complex. The implications are far-reaching, encompassing a wide range of fields that all rely on the efficient solutions of complicated problems in potential theory.

4.2 Neumann Problem

The Neumann problem, unlike its Dirichlet counterpart, doesn't prescribe values directly on the boundary; instead, it specifies the normal derivative. Imagine a heat-conducting plate: the Dirichlet problem fixes the temperature at the edges, while the Neumann problem dictates the rate of heat flow across those edges. This subtle difference leads to a fascinatingly different solution approach using conformal mappings. We're not dealing with direct function values but with their derivatives, a shift that requires careful consideration of the mapping's impact on the normal vector. The beauty lies in how conformal mappings, preserving angles, directly translate the normal derivative information across transformations.

Successfully solving a Neumann problem using conformal mapping often involves transforming the complex geometry into a simpler domain, like the upper half-plane. This transformation is typically a Schwarz-Christoffel mapping if we are dealing with polygonal boundaries. Then, we solve the simpler problem in the transformed domain, and employing the inverse mapping gives us the solution in the original geometry. Remember, the normal derivative changes under transformation, necessitating a careful application of the chain rule, along with the properties of the conformal map itself, specifically the Jacobian determinant. It's a mathematically rigorous process, not for the faint of heart, but the elegance and power of the technique are undeniably appealing.

The challenge, however, lies in determining the appropriate conformal mapping. Finding the correct transformation often requires a deep understanding of the geometry of the problem and the properties of the different conformal maps we have at our disposal. There's no one-size-fits-all solution. Careful analysis of the boundary conditions is crucial; each problem presents unique challenges and demands a tailored approach. Furthermore, the transformed problem itself may not be trivial to solve. For complicated geometries, numerical methods are often necessary to approximate the solution to the simpler problem in the transformed domain. The combination of analytical insight and numerical computation is frequently crucial in tackling real-world Neumann problems.

Consider a classic example: finding the electrostatic potential around a charged conductor with a specified surface charge density. The surface charge density directly relates to the normal derivative of the potential, positioning this as a classic Neumann problem. Here, a conformal mapping can simplify the conductor's shape into something manageable, such as a circle or a half-plane. The solution then becomes a much more tractable Laplace equation problem, and through the inverse mapping, we achieve a solution representing the original conductor's potential field. Remember that the accuracy of your final answer hinges on the careful execution of each step.

Numerical techniques prove invaluable when analytical solutions remain elusive. Boundary element methods, for example, offer a robust approach to handle intricate geometries that defy simple transformations. This approach involves discretizing the boundary, representing the solution as a summation of elementary contributions from each boundary segment. This method can be computationally intensive but provides a flexible and powerful tool for tackling problems beyond the reach of analytical solutions. This approach, while complex, offers a path towards solutions for problems with irregular or highly detailed boundaries that are often encountered in practical applications.

One point often overlooked: the constant of integration. Unlike the Dirichlet problem, where the boundary conditions uniquely determine the solution, the Neumann problem often introduces an undetermined constant. This stems from the fact that the normal derivative provides only information about the change in the function, not its absolute value. Physically, this reflects the freedom to shift the potential by a constant without altering the field, representing a choice of reference point. This constant necessitates additional physical constraints or considerations to obtain a unique and physically meaningful solution. It's a vital point to remember when applying these methods in practice.

In summary, tackling the Neumann problem using conformal mapping combines the elegance of analytical transformations with the robustness of numerical methods. It demands a strong grasp of complex analysis, careful consideration of boundary conditions, and a judicious choice of mapping. While the process might seem intricate, the reward is the ability to solve otherwise intractable problems in various fields, from fluid mechanics to electrostatics, showcasing the practical power and utility of conformal mappings in the world of physics and engineering. The key lies in choosing the right tools for the job, mastering the mathematical techniques, and paying attention to subtle details that can often make or break the solution. Ultimately, perseverance and attention to detail are essential to succeed with this fascinating method.

5 Applications in Fluid Dynamics

Conformal mapping provides a powerful tool for simplifying the analysis of fluid flow, particularly in scenarios involving complex geometries. Understanding potential flow, a cornerstone of fluid dynamics, is significantly enhanced through the use of these transformations. By mapping a complex flow domain onto a simpler one, such as the upper half-plane, we can solve the governing equations more easily and then map the solution back to the original domain to understand the fluid behavior. This approach bypasses the complexities of directly solving the equations within the intricate geometry of the original problem, leading to significant analytical simplification. This method is particularly useful when dealing with problems exhibiting irrotational and incompressible flow characteristics, common assumptions in many fluid dynamics applications.

The application of conformal mappings extends to the intriguing field of aerofoil theory, a pivotal area in aerospace engineering. The design and analysis of aircraft wings heavily rely on accurate prediction of air flow around these structures. The shape of an aerofoil, typically complex, is often mapped onto a simpler shape, like a circle, using conformal transformations. This allows for easier calculation of the pressure distribution around the wing using complex potential functions. Once this simplified analysis is complete, we transform the results back to the original aerofoil shape, giving us a robust understanding of lift generation and other critical aerodynamic parameters. This methodology enables efficient design of aerofoils with enhanced performance characteristics, minimizing drag and maximizing lift.

Beyond aerofoil theory, the versatility of conformal mapping extends to numerous other applications within fluid dynamics. Consider the analysis of flow around obstacles in a channel. The irregular geometry of the obstacle can present significant challenges in direct computational fluid dynamics (CFD) simulations. However, by skillfully applying conformal mapping techniques, we can simplify the computational domain, enabling more efficient and accurate numerical solutions. Similarly, problems concerning fluid flow in porous media, which involve complex network geometries, benefit immensely from the simplification provided by conformal mappings. The transformation can also aid in the study of free surface flows and flows in pipes with varying cross-sections, streamlining the analytical or numerical treatment of these complex phenomena. The efficacy of these mappings lies in their ability to transform complex boundaries into simpler, more tractable ones, opening the door to more straightforward analytical solutions or efficient numerical simulations. Researchers often use this technique for modeling groundwater flow through heterogeneous aquifers or analyzing blood flow in arteries with irregular shapes.

Consider the problem of flow past a cylinder. Directly calculating the velocity and pressure fields requires dealing with the curved boundary of the cylinder. Conformal mapping allows us to map this onto a simpler geometry, like a flat plate, where the solution is substantially easier to obtain. For instance, we can use a Joukowski transformation to map a circle onto an aerofoil shape. This simplified representation simplifies calculations immensely. Solving for the flow around the circle, then mapping the solution back to the aerofoil, provides a close approximation to the actual flow. This method significantly reduces the computational effort and complexity of evaluating the flow characteristics. This is particularly useful when dealing with flows exhibiting separation or other complex flow behavior that is difficult to simulate directly. The mapped solution then provides an approximate but practically useful result which can be checked against more computationally intensive methods.

Furthermore, beyond the already discussed applications, there are myriad other circumstances where this methodology is extremely useful. For instance, using conformal mappings to model the flow of fluids through complex networks, such as those found in microfluidic devices, is a key area. Likewise, the study of lubrication problems, where the geometry of the lubricating surfaces can be quite complex, is greatly simplified by employing conformal mappings. These transformations help researchers understand the pressure distribution and flow behavior in complex geometries, enabling improvements in lubrication design. This has implications in many engineering fields, from automotive design to advanced manufacturing processes. Conformal mapping thus provides an invaluable tool to analyze complicated flows, contributing to numerous critical advancements. It streamlines simulations, simplifies the modeling of various flow profiles, and assists in gaining valuable insights into fluid behavior in challenging geometries. Its applications span from the design of optimized aerofoils to the development of novel microfluidic devices.

5.1 Potential Flow

Potential flow, a cornerstone of fluid dynamics, finds elegant solutions through the power of conformal mapping. Imagine a complex flow pattern around an oddly shaped object–analyzing it directly is a daunting task. However, a conformal map can transform this intricate geometry into a simpler one, like a circle or a straight line, where the governing equations become significantly easier to solve. Once we obtain the solution in the transformed domain, the inverse transformation brings us back to the original problem's solution, providing valuable insights into the flow's velocity, pressure, and other crucial properties. This technique not only simplifies calculations but also provides a deeper understanding of the flow's behavior.

The application of conformal mapping to potential flow begins with recognizing that such flows are irrotational and incompressible. These conditions translate into Laplace's equation for the velocity potential, a fundamental equation in fluid mechanics. This equation, which describes the relationship between the potential function and the spatial coordinates, can be challenging to solve analytically for complex geometries. This is where the magic of conformal mapping truly shines.

Consider a problem involving flow around an airfoil, a critical aspect of aerodynamics. The airfoil's curved shape presents significant analytical difficulties. Using a Joukowski transformation, a specific type of conformal map, we can transform the airfoil's complex shape into a simpler circle. Solving Laplace's equation for the flow around the circle is considerably simpler. Afterward, the inverse transformation gives the solution for the original airfoil geometry, revealing pressure distribution, lift, and drag. The accuracy and efficiency gained from this approach are substantial. This showcases the power of conformal mappings in handling complex geometries where direct solutions are intractable.

Furthermore, conformal mapping's effectiveness extends beyond simplified geometries. In cases with multiple objects interacting, for example, multiple airfoils in a cascade or a flow around a structure with multiple components, direct computation can become excessively complex. Conformal mapping provides a streamlined solution. By transforming the geometry to a more manageable configuration, the interactions can be analyzed efficiently. This reduces computational cost while maintaining high accuracy. The resulting solutions offer valuable insights into the intricate interplay of fluid flows in these complex systems, crucial for applications such as designing efficient turbine blades or understanding fluid flow in biological systems.

The practical applications of this methodology are wide-ranging. In hydraulic engineering, for instance, analyzing the flow around bridge piers or dam structures is significantly simplified using conformal mapping. The irregular shapes of these structures pose a challenge for traditional analytical methods, but a clever choice of transformation can convert the problem to a more tractable form. The resulting solutions help optimize designs, minimizing turbulence and structural stress, leading to more efficient and safer infrastructure.

Beyond the purely analytical advantages, the visualization aspect of conformal mapping offers valuable insights. By plotting the flow field in the transformed domain, often a much simpler and more intuitive geometry, one can easily understand the overall flow pattern. This visualization helps identify regions of high velocity, stagnation points, and other critical features. The subsequent transformation back to the original domain retains the qualitative understanding gained from the transformed visualization, allowing for a more intuitive approach to understanding complex flow phenomena.

The ability to handle more complex scenarios, involving multiple interacting bodies or objects of arbitrary shapes, makes conformal mapping a robust tool in advanced fluid dynamics. For instance, understanding the flow around clusters of cylinders, as found in heat exchanger designs, significantly benefits from this technique. Each cylinder individually presents a challenge, but their interactions are vastly more complicated to handle directly. However, by a suitable conformal mapping, the problem can be tackled with significant efficiency and accuracy. The applications extend into diverse engineering disciplines, from aerospace to biomedical engineering, making conformal mapping an invaluable asset for modern engineering practice. Careful consideration of the appropriate transformation, based on the specific geometry and flow characteristics, remains crucial for successful implementation.

5.2 Aerofoil Theory

The elegance of conformal mapping truly shines when applied to aerofoil theory. Imagine the complex, curved shape of an aircraft wing; analyzing its airflow directly is a daunting task. However, through conformal mapping, we can transform this intricate geometry into a much simpler one—a circle, for instance—where calculations become significantly more manageable. This transformation, preserving angles, allows us to study the flow around the simpler shape and then inversely map the results back to the original aerofoil, revealing crucial insights into lift generation and pressure distribution. The power lies in the ability to solve a mathematically tractable problem in a transformed domain and then effortlessly translate the solution back to the real-world scenario.

Consider Joukowski's transformation, a classic example. This transformation, using a specific Möbius transformation followed by an inversion, maps a circle in the complex plane onto an aerofoil-like shape. By meticulously choosing the parameters of this transformation, we can effectively control the shape and characteristics of the resulting aerofoil. This allows for a systematic exploration of the relationship between the aerofoil's geometry and its aerodynamic performance. Analyzing the flow around the original circle, a problem readily solvable using elementary techniques, allows us to predict the lift and drag coefficients of the transformed aerofoil with remarkable accuracy.

The process isn't merely theoretical; it's deeply practical. Engineers frequently use these methods in the design of aircraft wings. By manipulating the parameters of the conformal map, designers can fine-tune the aerofoil shape, optimizing for maximum lift while minimizing drag. This iterative process, guided by the principles of conformal mapping, leads to the creation of efficient and reliable aircraft wings. The computational speed and efficiency offered by this approach are invaluable in the modern aerospace industry, enabling rapid prototyping and analysis.

Furthermore, the insights gained from conformal mapping extend beyond the basic aerofoil shape. More complex geometries, incorporating flaps, slats, and other control surfaces, can be tackled using more sophisticated transformations, although the level of complexity naturally increases. Nevertheless, the underlying principle remains the same: the transformation of a complex problem into a simpler, more easily solvable one, using the power of conformal mapping to bridge the gap between the mathematical model and the physical reality. The ability to accurately predict the flow field around such intricate structures is crucial for ensuring the safety and efficiency of aircraft.

Beyond the design phase, conformal mapping continues to play a vital role in aerodynamics. Understanding the flow around an aerofoil under various flight conditions, including high angles of attack, requires sophisticated modeling. Conformal mapping provides a crucial framework for solving the boundary value problem in these instances. By employing techniques like the Schwarz-Christoffel transformation, researchers can accurately simulate the airflow, even under extreme conditions, providing critical data for improved flight control and stability. This helps prevent potentially dangerous aerodynamic stalls, thereby enhancing safety standards.

Indeed, a profound understanding of conformal mapping is not just helpful; it is essential for advanced study in aerofoil design and analysis. While numerical methods have become increasingly sophisticated, the theoretical foundation provided by conformal mapping remains indispensable. It provides a powerful framework for understanding the complex interplay between geometry and aerodynamics, enabling us to design more efficient, safer, and more reliable aircraft. The principles outlined here aren't just applicable to aerofoils; they extend to a wide variety of applications in fluid dynamics, electrostatics, and other branches of physics and engineering, solidifying its place as a cornerstone of applied mathematics.

5.3 Other Applications

Beyond aerofoils, conformal mapping finds significant use in various fluid dynamics scenarios. Consider groundwater flow, where aquifers often possess irregular boundaries. Mapping these complex shapes to simpler geometries, like a circle or half-plane, dramatically simplifies the mathematical modelling of fluid flow through porous media. This allows engineers to predict water table heights and assess the impact of pumping wells with far greater accuracy than traditional numerical methods, leading to improved resource management and sustainable water extraction practices. The efficiency gains are substantial, especially in large-scale projects involving complex geological formations. Furthermore, the application extends to analyzing seepage beneath dams, ensuring structural integrity and minimizing the risk of failure.

The design of efficient heat exchangers hinges on optimizing the flow of fluids within intricate channels. Conformal mapping provides a powerful tool for designing and analyzing these systems. By transforming the complex geometry of the heat exchanger channels into simpler shapes, engineers can predict the heat transfer rates and optimize the design for maximum efficiency. This is particularly useful in applications where high heat transfer rates are required, such as in power generation plants or chemical processing industries. The enhanced design allows for smaller, lighter, and more efficient heat exchangers, reducing manufacturing costs and overall energy consumption. It's an essential tool in optimizing thermal management.

Hydrodynamic lubrication, the process by which a thin film of fluid separates two moving surfaces, is another area where conformal mapping shines. In traditional approaches, solving the Reynolds equation for the pressure distribution in the lubricant film can be computationally intensive for intricate bearing geometries. However, using conformal mapping to transform the complex bearing geometry into a simpler one significantly simplifies the problem, offering computationally efficient solutions. This leads to faster and more accurate design iterations, resulting in the development of more durable and efficient bearings across numerous industries, from automotive manufacturing to aerospace engineering.

Electrochemical systems, particularly those involving electrodeposition or corrosion, often feature complex geometries. Predicting the current distribution and potential fields within these systems is vital for optimizing process efficiency and controlling the properties of the resulting deposits. Conformal mapping proves indispensable in tackling this challenge. By mapping the irregular shapes of the electrodes to simpler geometries, engineers can more readily solve Laplace's equation for the potential distribution and determine the current density at various points on the electrode surfaces. This improved understanding leads to better control over the deposition process, producing more uniform coatings and reducing defects, with immediate benefits in applications ranging from manufacturing of integrated circuits to protective coatings on metals.

Oceanographic studies frequently involve irregular coastlines, making numerical modelling of ocean currents, wave propagation, and tidal effects computationally demanding. Conformal mapping offers a pathway to streamline these analyses. By mapping the complex coastline to a simpler domain, researchers can create models that are both accurate and computationally efficient, significantly reducing computational time and resources. The resulting improved understanding of ocean dynamics enables more accurate predictions of coastal erosion, wave action and the impact of climate change on coastal environments. This allows for better planning for coastal protection and efficient management of marine resources. This method is proving increasingly vital in regions with highly irregular coastal features.

6 Applications in Electrostatics

Electrostatics, the study of stationary electric charges, often presents complex geometries that challenge direct analytical solutions. However, conformal mapping offers an elegant and powerful approach to simplify these problems. By transforming a complex geometry into a simpler, often canonical, shape, we can solve Laplace's equation–the governing equation for electrostatic potentials–more readily. This transformation significantly simplifies capacitance calculations and electric field computations, allowing for accurate and efficient solutions in scenarios that would otherwise be intractable.

Consider the classic problem of determining the capacitance of a capacitor with unusual electrode shapes. Direct application of Gauss's law might prove incredibly difficult, potentially involving complex integrations over oddly-shaped surfaces. Conformal mapping changes this. By mapping the complex geometry of the capacitor electrodes onto a simpler geometry, like parallel plates, we effectively transform a complex problem into a far simpler one where the capacitance is readily calculable. This transformation doesn't alter the physical properties of the system; it simply allows us to work in a more convenient mathematical space, making calculations much more straightforward and computationally less expensive.

The electric field itself, a vector quantity, also benefits from this approach. While the electric field's direction and magnitude are initially challenging to ascertain for a complex electrode configuration, conformal mapping offers a systematic route to determining these parameters. After transforming the system's geometry into a simpler one, we calculate the electric field in the simpler geometry, a much easier task. We then use the properties of the conformal mapping to transform this simpler electric field back to the original, more complex geometry, obtaining a complete description of the electric field without the computational burdens of the direct approach. This is significantly important in high-precision simulations.

Let's illustrate this with an example. Imagine a capacitor formed by two elliptical electrodes. Determining the capacitance directly is notoriously challenging. However, we can use a Joukowski transformation, a specific conformal map, to transform these ellipses into two parallel plates. The capacitance of parallel plates is a well-known and easily calculable quantity. The beauty of this approach lies in the fact that the capacitance calculated for the parallel plates corresponds directly to the capacitance of the original elliptical capacitor. The conformal transformation facilitates this equivalent representation.

This method transcends simple capacitor designs. The use of conformal mapping extends to far more complex electrostatic problems, such as those involving irregularly shaped conductors or conductors in close proximity. In these instances, the complexity of the geometry may make direct computation infeasible or computationally expensive. Conformal mappings allow us to sidestep these difficulties. We find simpler, equivalent geometries where the electrostatic solutions are trivial. Then, through the application of the inverse conformal transformation, we accurately retrieve the solutions in the initial complex geometry. The resulting solutions are typically accurate and insightful, even for exceptionally challenging geometries. This is particularly useful when analyzing the behavior of electrostatic devices with irregular geometries.

The application of conformal mapping to electrostatics isn't merely theoretical; it has significant practical implications in the design and analysis of electronic components. Microelectronic devices, for instance, often incorporate components with intricate shapes and tight tolerances. Simulating the behavior of these devices using conformal mapping provides efficient and accurate results, aiding engineers in optimizing designs. This efficiency is crucial in a field driven by rapid technological advances, where quick, reliable simulations are necessary for the development of modern electronics. It's an invaluable tool in the hands of engineers.

Further exploration into specific conformal maps relevant to electrostatics, such as the Schwarz-Christoffel transformation for polygonal electrodes, will reveal further powerful applications. Moreover, understanding how the mapping impacts both the potential and field distributions allows for a comprehensive analysis that goes beyond simply calculating capacitance. This deep understanding fosters innovation.

In conclusion, conformal mapping provides a sophisticated and practical toolset for navigating the complexities of electrostatics. By transforming intricate geometries into simpler equivalents, we gain efficient and precise solutions to problems otherwise intractable using traditional methods. This elegance and practicality make conformal mapping an essential technique for anyone working in the field of electrostatics, allowing for significant improvements in both analytical and numerical analyses, directly influencing the design and optimization of a wide range of electronic systems. The power of this approach lies not just in its elegance, but in its practical application, transforming challenging calculations into manageable ones, contributing significantly to advancement in the realm of electrical engineering and applied physics. The ability to quickly and accurately model complex electrostatic systems offers a significant advantage in research and development.

6.1 Laplace's Equation

Laplace's equation, $^2V = 0$, is a cornerstone of electrostatics, describing the potential V in a charge-free region. Its elegance lies in its simplicity, yet its solutions unlock the secrets of electric fields within complex geometries, a challenge easily tackled using conformal mappings. Understanding this equation is crucial for anyone working with electric fields or similar phenomena governed by similar mathematical structures.

Conformal mappings provide a powerful tool for solving Laplace's equation in situations where direct solution proves intractable. The core idea is to transform a complex geometry into a simpler one, solve the equation in the simpler domain, and then map the solution back to the original geometry. This process leverages the fact that Laplace's equation remains invariant under conformal transformations; that is, if a function satisfies Laplace's equation in one domain, its conformal map satisfies it in the transformed domain as well.

Imagine you're tasked with finding the electric potential around a weirdly shaped conductor. A direct approach might involve solving a cumbersome partial differential equation. Instead, we can use a conformal mapping to transform this irregular shape into a simpler one, like a circle or a half-plane, for which the solution to Laplace's equation is readily available. This transformation doesn't alter the underlying physics; it just provides a more convenient mathematical framework. Once the solution is obtained in the simpler domain, the inverse conformal mapping accurately recovers the potential distribution in the original, complex geometry.

Consider the problem of finding the capacitance of a capacitor with non-standard plate shapes. This calculation is essential for numerous electronic devices, as capacitance directly affects the performance of circuits. A direct calculation from the geometry is often impossible without resorting to complex numerical methods. Employing conformal mappings, however, offers a streamlined approach. By mapping the capacitor's plates to simpler shapes, for instance, concentric circles or parallel plates, we can utilize known capacitance formulas. This transformed solution, upon mapping back, gives us a highly accurate capacitance value for the original, oddly shaped capacitor.

One particularly useful conformal map for solving Laplace's equation is the Joukowski transformation, often employed in aerodynamics but equally applicable to electrostatics. This transformation, by mapping circles to airfoils, allows us to analyze the field around complex airfoil shapes. It simplifies the calculation of the charge distribution and the electric field strength around the airfoil, offering crucial insights into the electrostatic behavior of these important components in electrical devices. The elegance of this method lies in its ability to handle shapes far more complex than the simpler cases often discussed in introductory texts. Furthermore, the solution to Laplace's equation obtained via conformal mapping provides not just the potential but also its gradient, which directly represents the electric field. Therefore, understanding the potential distribution allows us to readily calculate the electric field strength at any point within the region of interest. This complete characterization of the electrostatic field is critical for a variety of applications, from designing efficient electrical components to simulating the behavior of complex electrostatic systems. This detailed understanding of both the potential and the field is one of the key strengths of this method.

The practical application of these techniques extends far beyond simple calculations. Consider the design of high-voltage equipment where understanding the field distribution is paramount to prevent electrical breakdown. The conformal mapping approach allows for precise simulation of the field, aiding in the design of more robust and efficient high-voltage insulators and components. Moreover, the method finds applicability in optimizing the design of microelectronic devices, especially concerning the reduction of parasitic capacitance, which impacts device performance.

Beyond electrostatics, Laplace's equation and conformal mapping techniques find applications in many fields. Heat transfer, fluid dynamics, and even certain aspects of quantum mechanics are governed by equations that are structurally similar to Laplace's equation. Hence, the problem-solving strategies and the conformal mapping methods discussed here have broader significance across many areas of physics and engineering. The adaptability of this method underscores its importance as a powerful mathematical tool.

In summary, the application of conformal mapping to solve Laplace's equation offers a powerful and efficient approach to tackle challenging electrostatic problems. By transforming complex geometries into simpler ones, we simplify the task of solving the equation and readily gain insights into the electric potential and field distributions. This method, while rooted in elegant mathematics, offers remarkably practical and actionable advice for solving real-world engineering and physics problems. Understanding and mastering this technique provides a considerable advantage to anyone involved in these fields. The ability to apply it effectively separates theoretical understanding from practical problem-solving capabilities.

6.2 Capacitance Calculations

Calculating capacitance using conformal mappings offers a powerful alternative to traditional methods, particularly for complex geometries where direct computation proves difficult. Consider a capacitor formed by two conductors with an unusual shape; finding its capacitance using standard formulas becomes nearly impossible. However, by employing a conformal mapping, we can transform this intricate shape into a simpler, more easily analyzed geometry, often a parallel plate configuration. This transformation simplifies the calculation significantly.

The process begins by identifying a suitable conformal map that transforms the original conductor geometry into a simpler, canonical form. This choice hinges heavily on the specific shape of the conductors; common maps like the Joukowski transformation or Schwarz-Christoffel transformation may be employed depending on the complexity of the conductors. The selection necessitates a thorough understanding of the properties of various conformal transformations and their ability to map specific geometric features. Careful consideration of the mapping's properties is crucial for accurate results.

Once the appropriate conformal map is chosen, we apply it to the conductor boundaries. This transformation alters the geometry while preserving the angular relationships between intersecting surfaces. The capacitance remains unchanged under this transformation; this principle is pivotal to the effectiveness of this method. After the transformation, you are left with a configuration where capacitance calculations are much more straightforward. You could potentially now apply elementary formulas from electrostatics.

After transforming the complex geometry into a simpler form, the capacitance calculation becomes more tractable. The transformed geometry might be as simple as parallel plates, simplifying the calculation. For instance, the capacitance of parallel plates is well-known and easily calculated using the formula $C = A/d$, where is the permittivity of the dielectric material between the plates, A is the area of the plates, and d is the distance between them. It's vital to remember that the units of measure are critical for obtaining a correct solution. Double check every parameter before proceeding to the next step.

The calculated capacitance in the transformed configuration, however, represents the capacitance of the original, complex geometry. This equivalence underscores the power of conformal mapping in solving problems that would otherwise be computationally intensive or even impossible. Therefore, the final result directly reflects the capacitance of the original, complex capacitor. This non-intuitive result is a direct consequence of the angle-preserving nature of conformal mappings.

Let's illustrate this with a specific example. Suppose we have a capacitor constructed from two circular conductors. A direct calculation of the capacitance for this configuration would involve complex integration. However, we can utilize a bilinear transformation to map the circular conductors onto parallel plates. This transformation simplifies the problem; we can easily apply the parallel plate capacitance formula. This formula only requires the area and distance between the plates, values readily obtained after the mapping.

However, the practicality of this method relies heavily on our ability to find an appropriate conformal transformation. For some complex geometries, finding the suitable transformation may be a significant challenge. Advanced techniques, numerical methods, or even software implementations might be needed. For example, finding the Schwarz-Christoffel transformation for a highly irregular polygon may require numerical techniques.

In cases where a readily available analytical solution isn't possible, numerical methods offer a viable approach. These numerical techniques discretize the problem, allowing for approximate solutions through iterative processes. Software tools and specialized algorithms are often essential in these cases, offering a practical solution to problems which may otherwise be intractable. Such methodologies offer powerful and effective solutions to the problem of capacitance calculation even for extremely irregular geometries.

Beyond its application in capacitance calculations, the principle of conformal mapping extends to other aspects of electrostatics. Analyzing electric fields in complex geometries follows a similar process. By transforming the geometry to a simpler one, the problem of computing electric fields becomes much more manageable. The fields can be determined by solving Laplace's equation in the transformed coordinate system, then transforming the solution back to the original coordinates. The result yields the electric field distribution within the original, complex geometry. This approach simplifies field computations significantly, especially for irregular electrode shapes that otherwise defy simple analytical solution. Keep in mind that accuracy depends greatly on the choice of mapping and the precision of calculations involved.

In summary, calculating capacitance using conformal mapping provides a powerful and elegant method for solving problems in electrostatics, particularly for complex geometries that are intractable by traditional methods. While the selection of appropriate conformal transformations and, in some cases, the necessity for numerical methods add layers of complexity, this approach offers a flexible and versatile pathway to accurately determine capacitance and electric fields in scenarios that might otherwise be impossible to solve. The elegance and practicality of this technique make it an indispensable tool in the physicist or engineer's repertoire. Mastering this approach enhances the problem-solving capabilities significantly.

6.3 Field Computations

Calculating electric fields in complex geometries can be challenging. However, conformal mapping offers a powerful technique to simplify these calculations, transforming intricate shapes into simpler ones where the field is easier to determine. This transformation hinges on the fact that

Laplace's equation, governing electrostatic potentials, remains invariant under conformal mappings. The process involves mapping the complex geometry onto a simpler domain, solving Laplace's equation in this simpler domain, and then using the inverse mapping to obtain the solution in the original geometry. This method drastically reduces the computational complexity involved in solving boundary value problems with complex boundary conditions. The elegance lies in its ability to convert a seemingly intractable problem into a readily solvable one, demonstrating the power and utility of conformal mapping techniques.

Consider, for example, the problem of determining the electric field distribution around a pair of charged parallel plates with an irregularly shaped cutout in one of the plates. A direct solution using numerical methods like finite element analysis would require significant computational resources and could be prone to numerical instability due to the irregular geometry. Using conformal mapping, we can transform the complex geometry of the plates with the cutout onto a simpler geometry, perhaps a pair of parallel plates without the cutout. Solving Laplace's equation for this simplified geometry is significantly easier and faster. The potential distribution in the simpler domain can then be transformed back to the original complex domain via the inverse conformal mapping, yielding the desired electric field distribution around the irregularly shaped plates. This provides a precise and efficient solution without the computational burdens associated with more traditional numerical techniques.

The choice of appropriate conformal mapping is crucial for the effectiveness of this approach. The mapping should ideally transform the complex geometry into a simpler, canonical shape like a rectangle, a circle, or the upper half-plane. This allows for the utilization of readily available solutions for Laplace's equation in these standard domains. Several types of conformal maps, such as the Möbius transformations, the logarithmic map, and the Schwarz-Christoffel transformation, might be appropriate depending on the shape and complexity of the original geometry. The Schwarz-Christoffel transformation, in particular, is a powerful tool for mapping polygons onto the upper half-plane, making it suitable for a wide range of problems involving polygonal electrodes or conductors. Careful consideration of the mapping's properties, including its singularity locations and branch cuts, is necessary to ensure the accuracy and validity of the resulting solution.

Practical implementation often involves numerical techniques, even after applying the conformal mapping. The mapping itself may be defined implicitly and require numerical methods for its evaluation. Moreover, even in the simplified domain, solving Laplace's equation might require numerical techniques such as finite difference or finite element methods. However, the computational cost is significantly reduced compared to tackling the original problem directly, particularly for complex geometries. The advantage comes from the pre-processing step, simplifying the geometry and allowing for faster and more stable numerical solutions. Software packages specializing in numerical conformal mapping can simplify this process, offering streamlined workflows for mapping generation, boundary condition implementation, and solution retrieval. The efficiency gains are particularly significant in problems involving iterative refinement or optimization.

Beyond the capacitance calculations discussed earlier, field computation via conformal mapping has broad applications within electrostatics. For instance, consider designing an electrode system for a high-voltage device. Optimizing electrode shape for minimal field concentration near the surface is crucial to prevent dielectric breakdown. Traditional design methods might involve costly trial-and-error procedures involving experimental prototyping. Utilizing conformal mapping allows for simulating different electrode geometries with relative ease, quickly identifying designs that minimize field concentration and hence improve device reliability. This approach significantly reduces design time and costs. Similarly, understanding the field distribution around charged particles within complex dielectric materials can be greatly simplified by employing conformal mappings. This is highly relevant in fields like material science, where understanding charge distribution at an atomic level is crucial. The ability to precisely calculate these fields opens up possibilities for optimizing material properties and device performance. In conclusion, conformal mapping provides a powerful computational framework enabling efficient and accurate calculations in electrostatics.

7 Advanced Topics & Further Reading

Riemann surfaces represent a significant advancement beyond the plane geometry we've explored thus far. Imagine extending the concept of a complex plane to encompass multiple sheets, seamlessly connected along branch cuts. This multi-sheeted structure allows for the representation of multi-valued functions, like the square root, in a consistent and geometrically meaningful way. Understanding Riemann surfaces requires a deeper grasp of topology and complex analysis, particularly concepts like covering spaces and branch points. Their study illuminates the inherent complexity of conformal mappings and opens up avenues for analyzing functions with multiple values and branch cuts. This expanded framework is vital for understanding advanced problems in complex analysis and their applications in various scientific fields. Further study might involve delving into the theory of covering maps and exploring the profound connections between Riemann surfaces and algebraic geometry. Think of visualizing a function's behavior not just on a single plane, but across interconnected sheets, revealing the hidden complexity and elegant structure often obscured in a single plane representation. Mastering this allows for a deeper comprehension of highly complex mathematical structures and their applications.

Discrete conformal maps provide a powerful computational approach to conformal mapping problems, especially where analytical solutions are intractable. Instead of working with continuous functions, these methods discretize the problem, representing the domain and mappings using a mesh of points. Algorithms such as circle packings or quasi-conformal mappings are employed to approximate the desired conformal map numerically. The accuracy of the approximation depends on the mesh resolution and the specific algorithm used. This numerical approach is critical for solving problems involving complex geometries or intricate boundary conditions, which often defy analytic solutions. They are essential tools in many applications, including image registration, surface parameterization in computer graphics, and the numerical solution of boundary value problems in physical systems. These methods allow one to handle intricate and challenging geometries, providing approximate solutions where precise analytic solutions are out of reach. Imagine the power of numerically approximating complex mappings to solve problems that are otherwise unsolvable using analytical methods—that's the utility of discrete conformal maps.

For those seeking to delve deeper into the fascinating world of conformal mappings, several excellent resources are available. Classical texts like Ahlfors' "Complex Analysis" provide a rigorous mathematical foundation. More applied treatments, focusing on specific applications such as fluid dynamics or electrostatics, often highlight the practical utility of these techniques. Nevanlinna's work offers insights into the deeper mathematical aspects, while treatises on numerical methods provide essential tools for solving complex problems computationally. Exploring these resources will lead to a more complete understanding of both the theoretical underpinnings and the practical applications of conformal mappings. Seek out books and papers focused on specific areas of interest; the depth of literature on conformal mappings is vast, spanning various mathematical and engineering disciplines. Remember, continuous learning is crucial; the field is constantly evolving, with new advancements and applications emerging regularly. Therefore, staying updated with the latest research is pivotal for those seriously pursuing this field.

Exploring the connections between conformal mapping and other areas of mathematics is also highly rewarding. For instance, the link between conformal maps and Riemann surfaces reveals the profound geometric structures underlying complex analysis. Further, the relationship to partial differential equations, particularly Laplace's equation, emphasizes the significant role of conformal mapping in solving boundary value problems in physics and engineering. The application of conformal mappings in diverse fields, ranging from fluid mechanics and electrostatics to computer graphics and image processing, highlights its versatility and enduring significance in modern science and technology. These cross-disciplinary applications demonstrate the power and breadth of conformal mapping as a valuable mathematical tool in many domains. The connections between conformal mapping and other mathematical fields and applications offer a wider perspective and a deeper understanding of its significance.

7.1 Riemann Surfaces

Riemann surfaces offer a powerful generalization of complex analysis, extending the familiar concepts of analytic functions to multi-sheeted surfaces. Imagine trying to define a square root function across the entire complex plane; a single value isn't sufficient, as there are always two possible square roots for any nonzero complex number. This ambiguity is precisely what Riemann surfaces elegantly resolve. They provide a geometric framework where functions that are multivalued in the complex plane become single-valued on these extended surfaces. This is accomplished by creating multiple "sheets" connected at branch points–points where the multiple values coalesce. Consider the simple example of the function w = z^(1/2). The Riemann surface for this function consists of two sheets, each representing one branch of the square root. These sheets are connected along a branch cut, typically chosen to be the negative real axis. Navigating along a path that encircles the origin and crosses the branch cut transitions you between the sheets, illustrating the multivalued nature of the square root in the complex plane but the single-valued nature on the Riemann surface.

The concept of analytic continuation finds a natural home on Riemann surfaces. Recall how analytic continuation extends an analytic function beyond its initial domain of definition. On a Riemann surface, analytic continuation is represented by moving smoothly across the surface. Branch cuts, those seemingly artificial boundaries in the complex plane, become natural boundaries on the surface. Therefore, Riemann surfaces clarify the process of analytic continuation and its inherent limitations. In essence, they provide a geometrical interpretation to a fundamental concept of complex analysis, transforming an abstract process into a visual journey across sheets. It's important to grasp this connection; it truly illuminates the profound implications of Riemann surfaces.

The topology of a Riemann surface dictates its properties and applications. Compact Riemann surfaces, for instance, possess a finite number of sheets and are incredibly important in many areas of mathematics and physics. They are classified by their genus, a topological invariant that essentially counts the number of "holes" in the surface. A sphere has genus zero, a torus has genus one, and so on. This genus plays a crucial role in determining the behavior of functions on the surface. Understanding this topological aspect helps to understand the overall structure and behavior of the functions defined on these surfaces. Moreover, the genus provides a way to classify different Riemann surfaces and their unique properties. The more complex the topology, the richer the mathematical structures that can be defined upon it.

Consider applications to algebraic geometry. Algebraic curves, defined by polynomial equations, can be represented as Riemann surfaces. The study of algebraic curves, therefore, is intrinsically linked to the study of Riemann surfaces. This connection leads to powerful techniques for analyzing the properties of these curves. For example, the genus of an algebraic curve determines the number of linearly independent holomorphic differentials on the corresponding Riemann surface–a crucial concept in the theory of algebraic curves. This illustrates the interplay between algebraic geometry and complex analysis, unified by the concept of Riemann surfaces. This is not merely a theoretical curiosity; it is a tool that provides deep insights into the structure of algebraic curves and their associated solutions.

Furthermore, Riemann surfaces appear in various physical contexts. In string theory, for instance, the worldsheet of a string is often modeled as a Riemann surface. The topology of the worldsheet determines the properties of the string and its interactions. Different topologies, represented by different Riemann surfaces, correspond to different types of string configurations and interactions. This highlights the importance of understanding the topological properties of Riemann surfaces in advanced physical theories. The study of Riemann surfaces, seemingly an abstract mathematical concept, finds deep connections within the framework of theoretical physics and mathematical modelling. From topology to string theory, the implications are far reaching and transformative. This connection between sophisticated mathematics and fundamental physics is rarely seen so plainly expressed.

Finally, numerical methods play a crucial role in working with Riemann surfaces, especially for more complex surfaces. Discretization techniques, often involving triangulations or other meshing methods, allow us to approximate the surface and perform computations on it. These numerical approaches are particularly essential for solving boundary value problems or simulating physical processes on these surfaces, allowing researchers to perform practical computations for problems not otherwise easily solved. Furthermore, the field of computational conformal geometry is actively developing new methods for efficient and accurate calculations on Riemann surfaces and their applications. This demonstrates the synergy between classical mathematical theory and modern computational techniques—a critical aspect of present-day scientific and engineering research. The applications are far-reaching and continue to evolve as computational power grows.

7.2 Discrete Conformal Maps

Discrete conformal maps offer a powerful approach to tackling conformal mapping problems in the digital realm. Unlike the continuous mappings discussed previously, these methods deal with discrete representations of surfaces and functions, making them ideal for computer implementation and analysis. They're particularly useful when dealing with complex geometries or situations where analytical solutions are unavailable. Imagine trying to map an irregularly shaped coastline–a continuous conformal map might be incredibly challenging to find, but a discrete approach allows for approximation using a mesh or grid.

The core idea behind discrete conformal maps rests on adapting the principles of angle preservation to a discrete setting. Instead of considering smooth functions, we work with graphs representing the surfaces. The vertices of the graph correspond to points on the surface, and the edges connect neighboring points. A discrete conformal map then seeks to maintain the angles between edges in a way that mirrors the behavior of a continuous conformal map. Various algorithms exist for constructing such maps, each with its own strengths and weaknesses. These algorithmic variations often involve iterative processes, refining the mapping until a suitable level of angle preservation is achieved.

One common approach involves using a quasi-conformal mapping. Quasi-conformal maps allow for slight distortions in angle preservation, providing a balance between accuracy and computational feasibility. This flexibility is critical when dealing with complex geometries or limitations in computational resources. While perfect angle preservation might be computationally expensive or even impossible in some cases, quasi-conformal maps offer a practical compromise, often yielding highly accurate results suitable for many engineering and scientific applications. Think of it like using a flexible approximation–sometimes close enough is good enough, especially when facing computational constraints.

The application of discrete conformal maps extends beyond theoretical mathematics. They find significant use in image processing and computer graphics. Imagine warping an image to fit a specific target shape, maintaining the original image's structure and proportions as much as possible. Discrete conformal maps provide the mathematical framework for achieving this distortion while minimizing unwanted artifacts. This warping, crucial in image registration and texture mapping, allows for seamless integration of images onto complex 3D models or surfaces. The ability to accurately distort an image while preserving its integrity is invaluable in many digital applications.

Furthermore, these mappings are instrumental in mesh generation. Creating high-quality meshes for numerical simulations often requires intricate mappings to handle complex geometries. Discrete conformal maps ensure the mesh elements are well-distributed, avoiding excessive skewing or distortion. This leads to increased accuracy and stability in numerical simulations, making them crucial in fields such as computational fluid dynamics and finite element analysis. A well-distributed mesh is the backbone of accurate numerical solutions, crucial in various computational simulations.

The choice of a specific discrete conformal mapping algorithm often depends on the nature of the problem and the available computational resources. Some algorithms are highly efficient but might only handle specific types of surfaces, while others are more versatile but computationally intensive. Understanding these trade-offs is essential for successfully implementing discrete conformal mapping techniques. The optimal approach often involves balancing the desire for accuracy and the constraints imposed by computational power and time.

Another critical consideration is error analysis. Since discrete conformal maps are approximations, understanding and quantifying the errors introduced by the discretization process is paramount. This involves developing strategies to estimate or bound the error and ensuring the errors remain within acceptable tolerances. Careful error analysis ensures that the results obtained using discrete conformal maps are trustworthy and reliable for use in practical applications. The goal is to make sure the approximation is sufficient for the task, not just aesthetically pleasing.

In conclusion, discrete conformal maps provide a powerful and practical set of tools for tackling conformal mapping problems in computational contexts. Their ability to handle complex geometries and the adaptability of various algorithms make them indispensable in fields ranging from image processing to numerical simulations. The careful consideration of algorithmic choices, error analysis, and computational resources is essential to ensuring the effective and reliable application of these methods. Understanding these techniques forms a crucial bridge between the theoretical elegance of conformal mappings and their practical implementation in the digital world. The future of this field will likely involve developing even more efficient and robust algorithms capable of handling increasingly complex geometries and diverse applications.

www.ingramcontent.com/pod-product-compliance
Lightning Source LLC
LaVergne TN
LVHW022018200926
848481LV00007B/2447